HAM RADIO
2020 For Elderly Technicians, Extras and General License

A Quick Reference to Obtaining License and Setting up Ham Radio

COLIN
SCHMITT

Copyright

Copyright©2020 Colin Schmitt

ISBN: 9798669212179

All rights reserved. No part of this book may be reproduced or used in any manner without the prior written permission of the copyright owner, except for the use of brief quotations in a book review.

While the advice and information in this book are believed to be true and accurate at the date of publication, neither the authors nor the editors nor the publisher can accept any legal responsibility for any errors or omissions that may be made. The publisher makes no warranty, express or implied, with respect to the material contained herein.

Printed on acid-free paper.

Contents

Copyright i

CHAPTER ONE 1

UNDERSTANDING THE HAM RADIO 1

Tuning in to hams radio 6

Electronic and technology aspect of ham 6

Social aspect of Ham 11

Making contacts with other hams................... 11

Joining the ham radio community 25

Differences between UHF and VHF 27

VHF – Very High Frequency............................ 28

UHF – Ultra High Frequency........................... 30

Building your Ham radio Equipment 33

CHAPTER TWO 35

GETTING MORE FROM HAM 35

Feedlines and Filter gadgetry in ham 35

Filters 36

Getting your Ham on the Air 38

How natural environment affects radio transmission 42

Joining ham clubs 44

Finding other hams...................................... 44

Radio clubs 45

Finding and choosing the right ham radio club 47

Improving your club participation 49

Getting involved in ham radio group.............. 51

Special purpose Ham organizations and clubs 58

Handi-Hams 60

AMSAT 61

Hamfests and Conventions........................... 63

Finding hamfests 64

Buying at hamfests 66

Finding conventions 67

Tips for ham radio operators 69

CHAPTER THREE 72

LICENSING THE HAM 72

Frequency assignment 73

Licensing classes 74

The Technician License 75

The General License 77

The Amateur Extra License 78

Grandfathered license classes 79

Taking the license test 80

Finding the right material for study 81

The Volunteer Licensing System 87

Volunteer Examiner Coordinator (VEC) 89

Finding a Test Session 90

Public and Private Exams 92

Understanding Call Signs 93

GETTING YOUR LICENSE 95

Completing your paperwork 95

Finding Your New Call Sign 97

CHAPTER FOUR 103

USING HAM IN EMERGENCY 103

ARES and RACES 106

Preparing for an Emergency 108

Tips to get ahead during emergency 110

Reporting an accident: Tips 110

CHAPTER FIVE... 113

MAINTAINING YOUR STATION 113

Tools for maintaining your equipment......... 113

Routine maintenance for stations 116

ABOUT AUTHOR 118

CHAPTER ONE

UNDERSTANDING THE HAM RADIO

The term "ham" has been in existence for more than a century. The term has been deployed by individuals or corporate bodies who, for one reason or another, enjoy being on air for recreational, educational or other technical reasons. A ham operator, also addressed as amateur radio operator, can build or personalize his own radio station after obtaining such license from relevant bodies. The ham operator essentially uses equipment from a ham radio station to initiate a two way communication with other ham operators, or even individuals within the community. The ham radio station will be able to operate on specific frequencies which have been assigned to the station by relevant communication agencies – such as the Federal Communication Commission (domiciled in the United States) and the International communication union (found in all countries of the world). Ham radio operators can create and use many types of radio stations, ranging from permanent ground stations, field stations (temporary), space stations and stations set up on mobile.

The widespread popularity of ham radio was spurred by the willingness of governments and regulators all over the world to assign a portion of the radio spectrum to interested citizens. Before you are assigned your radio frequency, you need to showcase your ability and your dedication to utilize the radio spectrum using the right guidelines. For people in the United States of America, the government body governing the assignment of the radio frequency is the Federal Comm-unications Commission. The Federal Communications Commission is also the same body that is dedicated to seeing to the license of broadcast stations, and to also ensure that radio users comply with the right guidelines. To operate a Ham radio, you must sit for and pass an exam – spanning through tests on different operation practices, governing regulations and electronic tests concerning radio operations.

Amateur radio operators are especially important, as they dedicate their times to provide emergency communications between communities and places during disasters. Disasters can be man-made or natural, and people might not have been aware of the spread. They also serve as volunteers during some special events such as sporting activities and other public shows. Essentially, amateur radio operators are useful in providing safety to the public. There is a law preventing ham operators from accepting monetary rewards from fans who are only swayed by their dedication and excellent communication services. But today, the law is not profound within the ham communities.

Aside from ham radios, there are other communication systems that can be used for person-person communications. But these forms of communication systems are usually restricted and cannot work beyond some specific frequencies. For instance, the system that processes cellular calls can only accommodate some specific number of subscribers at a time. You can only imagine if a disaster occurs and they need to send information across to many people at the same time – then this type of communication system is not particularly reliable for that purpose. Ham radio, on the other hand, deploys a continuous range of frequencies – and sending information continuously cannot be a problem. Many ham radios are found operating from frequency above the **AM** band (Amplitude modulated frequency) to just above the high frequency citizen band. The **AM** band operates between **535 to 1605 kHz** while the **citizen** bands are found between **26.9 to 27.4 MHz.** Long distance communications during the daylight are better with **15 to 27MHz** band, while the **1.6 to 15 MHz** bands are good for long distance communications at night. These bands (**AM**

and citizens band) are often referred to as **short-wave bands.** The frequency used by your TV set at home is not the same with the frequency used by the Ham radio. The TV set in your house has frequency referred to, literally, as line-of-sight frequency and cannot travel more than 50 miles. The ham radio (**short wave band)** bounces off the ionosphere from the transmitter to the antenna of the receiver. The ionosphere reflects radio waves easily, and these bouncing radio waves improve communication over long distance. This implies that the ham radio wave travels faster than your TV set. It was understood from basic science that "the higher the frequency, the shorter the wavelength."

There have been some improvements in the ham radio today – compared to how it was some decades ago – with ham radios now using the **Amateur radio satellite.** With the **Amateur radio satellite,** radio operators can now use their hand-held radio to communicate over the amateur radio satellite.

Tuning in to hams radio

There are many reasons why people decide to venture in or to use hams radio. These reasons range from people who are just interested in the fun side of ham radio to those who have interest in understanding the basic aspect of radio electronics. We can have the technology and operation aspect of ham radio and also the social aspect (using ham for fun).

Electronic and technology aspect of ham

The ham radio deploys electronic signals, and these signals must be transmitted for sound to be heard. The transmission and reception of radio signals has to do with a lot of electronics. As soon as you have decided to get started with ham radio, you will be opened to a lot of things from electronics (things that constitute the bulk of ham) to some advanced radio-frequency methods. As a ham operator, you may opt to build and customize your own ham gadgets or to build a station from materials that have been ready-made from the factory. Most materials you will require – no matter which route you want to take- are available on the web and in stores. All that you only need to do is to shop online for equipment that will enable you to get the most out of ham. Ham radio operators have a do-it-yourself slogan, popularly called home-brewing, and often help each other in times of need – especially to build and organize their stations. One beautiful thing about ham radio operators is that they have the liberty to make their own software and deploy the internet to make a real hybrid system. Some years back, ham operators were able to come together to develop packet radio using data transmission protocol in computer networks to ham radio link. The Automatic Position Reporting System (APRS) was created by combining GPS radiolocation technology together with the web and ham mobile radio. The most common technologies by which ham radio operators talk to each other is the Voice and Morse code communications. But nowadays, computer-based digital communications and operations are gaining momentum. Today, many home station communications now deploy a hybrid of computer and radio. The software-defined radio (SDR) technology is currently being explored by many of the newer radios which makes the future of ham looks more promising. The software-defined radio (SDR) technology enables reconstitution of the electronics that processes the radio signals using software control. The computers and the high-end equipment being used by ham radio operators are just a few out of the many tools that make up the ham radio components. The ham radio operators use antennas and propagation, which form the bulk of the means through which radio signals move around from one place to another. Hams radio operators are particularly interested in sunspot, solar cycles and their significant effect on radio transmission. For instance, if the sunspots are active, there will be more solar flares which will create an increased geomagnetic activity in the earth. At maximum sunspots, there might be a possible disruption in power grids and radio transmission. For the ham radio operators, weather can be both advantageous and disadvantageous. Weather can be advantageous for hams because it can create fronts along which radio can travel long distances. At the same time, weather can be a hindrance to hams; for instance during rain and snow. Rain (water) is a conductor of electricity and can absorb radio waves which consequently reduces signal strength. In fact, if the radio wave travels through trees, the water content in the tree can reduce radio reception. The earth temperature also affects short wave transmission which consequently reduces radio reception; this can even be compounded in the presence of solar flares. Antennas, through which signals are launched to take advantage of all the sound propagation and receptions, initiates a level-playground for the ham station builder and even ham experimenters who are just coming onboard to gain ground experiences. With antenna experimentation, there are many things for ham radio operators and beginners to learn. With more and more hams springing up everyday across the United States, the antenna systems have seen many refinements in terms of designs and arts. This, even, is useful to people using antennas and radio signals for other purposes aside from ham radio. Antenna systems can be just some small patches of printed circuit board gadgets or some multiple ground towers decorated with many large rotating arrays and designs. All that you will require to come up with your own user antenna are just some few wires, a feedline and a soldering iron. Ham radio operators also deploy radio technology in tasks such as radio control, meteorology and model rocketry. Ham radio operators have special features for radio control operation in the 6-meter band very far away from the usual crowded unlicensed radio control frequencies. Small ham radio video transmitters are sometimes used in rockets, balloons and some model aircrafts; to access pictures from a height of a few thousand feet. Data from ham are also useful, especially, in astronomy, aviation, car racing and rallies amidst many other uses.

Social aspect of Ham

Making contacts with other hams

If you quickly tune your radio across the ham band (more on how to do this later), you will be able to hear hams (radio operators) making contacts – from simple and harmless chatting to on-air meetings. Note that in this guide, you may come across **hams** being used in place of **ham radio operators;** they both mean the same thing and can be used interchangeably. See below the various ways hams communicate;

1. **Ragchewing or Ragchew:** Hams make contact with each other. This type of communication between hams is commonly referred to as **contact.** The contact, literally, means the exchange of information between two ham stations. The contact starts when one operator makes an initial call to another ham station. The receiver amateur radio station gives out a response from their own amateur station and a signal report of the contact can be monitored. The contact is usually referred to by the **Q CODE.** The **Q CODE** is usually some collections of three letters code which stand for various meanings in the amateur radio community. Different organizations have their own codes which have meaning to them. These codes may not exactly have any meaning to someone outside of the discipline. The **Q CODE** must always start with letter **Q** and they are normally deployed to shorten lengthy phrases

and pass quick communication. For instance, if a ham says **QRA**, he is literally asking the other ham in contact what the name of his station is. Two or more stations that have already established contacts with each other can be said to have **worked** each other. An amateur radio operator can also say that he has **worked** a state, a community or a country.

The slang *ragchew or ragchewing* is often deployed in the ham community to mean an informal chit-chat between two hams. When the operators are in person-person contact, it is called an "eyeball QSO." The ATNO (All-time New One) is used to mean a type of connection between a ham and a radio station that have never worked together before on any platform. Many hams will usually send a "can you acknowledge receipt" shortened as QSL card that they have once worked with. QSL cards can be kept and used as an evidence of contact between two stations.

Check below a list of some important **Q CODE** to get familiar with how **q codes** work in hams;

- **QRG:** hams say this to ask for their exact frequency or that of another ham. It is a short form for "will you tell me my exact frequency or that of …."
- **QRL:** to ask another hams if they are busy or not. It is a short form for 'are you busy?
- **QRS:** to ask if to send more slowly. It is a short form of "shall I send more slowly?
- **QRQ:** to ask if to send faster. A short form for "shall I send faster?
- **QRV:** for "are you ready?
- **QRZ:** meaning "who is calling?

Note that responses to these radiotelegraph queries vary depending on the meaning of the code (what was asked). Usually, the letter "**R**" is usually sent when the **Q CODE** only needs to be acknowledged and no specific response is required. The letter "**R**" denotes "**Roger**" which means the code has been "received correctly."

Hams can ragchew across continents or towns. This means that you don't need to know the other hams before you set up communication with them. No hobby can be friendlier than ham radio: no discrimination – just call and start talking.

- **Knowing where to chew the rag:** It is not enough to know how to set up contact between you and other hams; but knowing the exact time and period you can chew the rag is also very expedient. When you (a ham operator) are looking to chew the rag with another ham, you should consider the following tips;

    o **HF Bands:** You can find contacts of many hams on a typical HF band. The HF band is the **high frequency** band denoting the range of frequencies between 3 and 30 MHz. The HF band can penetrate through the earth with little to zero interaction. You will find the ragchewers (contacts) mixed with some long distance contacts (DX) at the low end of the HF band. You also see the continuous wave (CW) at the lower frequency spectrum of the High Frequency bands. The faster radio operator will be at the bottom of the band, and the code speed will start dropping slowly as you tune higher. You find the ragchewers mixed in with the long-distance (DX) contacts at the low end of the band.

You may think that ragchewers are knocked off from all sides, but this is not always so. There are always ragchewing contact taking place all the time, so for ragchewers, they seem to occupy almost all spare of bands. If there is an impending disaster or an active disaster ongoing in a country or a specific location, some major ongoing events or period of big games or contest; the band may appear to be too full or too occupied to get anything done at all. In a situation like this, you can try another band as you are likely to find many available bands. This is one advantage of hams; giving you the opportunity to stay connected at all time – there are many nice frequencies to always get started anyways.

- **VHF** and **UHF Bands:** Ragchewers are usually found in the repeater section on the UHF and the VHF bands. The weak signal part of these bands usually features an SSB, CW or some other incorporated digital systems as wide-open spaces. If you want to tune in with other hams, you can scan the wide-open spaces far away from the repeater channel to have access to a local group. Once you can see the frequency, just hop in and say hello hams. The weak signals are located at the bottom part of the UHF/VHF band. The reason they are called weak signals is because contacts via SSB and CW are possible with weak signals; and there is no problem with sound propagation using these weak signals. Most of these weaker signals are especially perfect for good contacts. When you incorporate digital systems, you can have a digital repeater network capable of establishing internet-like modes for persons with similar intention; to connect. For instance, the WIRES-X network features "rooms" and the digital mobile radio (DMR) system features "talk groups." These two features function similarly with the analog frequency modulation (FM) calling frequencies providing a better way to make contact with other stations with little no searching.

- **Knowing when to chew the rag:** There are good times and bad times for ragchewing. Let us say you have been having signals hitting on you from various sources, and you are thinking whether this is a good time to engage in ragchewing with other hams. Well, there are no hard and fast rules about this. But you might probably need to check the time and days that are most convenient for your contacts. Weekdays might be perfect timing for ragchewing – particularly the daylight period – when other hams are at work. You need to confirm if the other ham is not engaged with other activities at the time you want to make contacts. The time discrepancy from countries to countries should also be taken into consideration; you don't want to disturb the other ham when he is sleeping or resting. Many ham operators also engaged in ham activities during weekends, though rare and only when some special events or activities are to be covered. But you can be rest assured that ham contacts are available every day of the week (weekdays to weekends). If you tune into one mode and it is busy, you can immediately tune into another mode. The World Administrative Radio Conference bands are always available for ragchewing and casual operations.

- **How to tell if a station is ready to ragchew or not:** If you have identified that you are actually ready to ragchew, how can you know if a station will like to ragchew with you? Well, one easiest route is to identify an ongoing ragchew session and then hop in. You can join the session or wait until one of the stations is logging out of the session and you can then place calls across to the other stations. You can also get familiar with some basic cues that can depict that a station is ready to ragchew with you. Such cues might include an observable relaxed voice tone. Also, look for a station that has consistent signal strength and one that is easy to copy. The best ragchew connections, usually, are the connections that are consistent enough to get you beyond the opening pleasantries – this is why you need to find a signal strength that is consistent enough. You can easily guess if a station is not looking forward to ragchewing when you hear a targeted call. For instance, when you hear "CQ Washington from A8BMI," what this means is that A8BMI has some information to pass across. CQ here means A8BMI is looking forward to communicating with the Washington ham operator. And once you are not on one of the targeted hams, you might not likely get in, and you should continue tuning to other stations. Also, when you see a call that has a lot of stations responding to it; this might

be another cue not to ragchew. This is because the station might be in another event at that moment. This means that you should continue till luck shines if you are really interested in ragchewing.
- Round table discussions are also good ways to get started with ragchewing. Roundtable discussions, essentially, involve contacts between more than two hams operating on a single frequency. Just like you and your friends meeting together for lunch. If only you can arrange such a meeting that only one of you gets to talk at a particular time, then that is a roundtable discussion.

Keep these things in mind while you're chewing the rag:
- **Initiate the call by starting with your basic information:** say your signal report, call sign, operator name, and your station location. Ragchew contacts can be set up between hams in different locations across the world. As soon as you discuss some basic information, you can spin off conversations in other directions. Talk

about other hobbies, families, and interesting topics- just any interesting topics. Avoid talking about politics and religious issues. Don't use abusive words with other hams.
- **Put an end to the conversation when you run out of ideas and things to discuss or when you are finding it difficult to maintain interesting contacts.**

2. **Networks:** This is another means, aside from ragchewing, which ham operators use to connect with each other. Nets are organized on-the-air meetings initiated for hams that have a similar interest or purpose. Some of the nets you can find can include;

Traffic nets: These are extensions of the North American system that takes text messages or traffic via ham radio. Hams operators come together to exchange or communicate messages, sometimes handling many such messages in a day. These messages range from the normal everyday chitchat or mundane to emergency health-and-welfare.

Emergency service nets: Most of the time, these networks literally meet for training and practice. When disasters or any other emergencies erupt, ham radio operators will organize themselves around these networks to provide very vital communications into and out of the affected areas until normal communications are restored.

Technical Service: These networks are just normal radio call-in programs where stations place calls across to ask specific questions or solve some problems. The network control station may help provide the answer, but more often than not, it is one of the listening stations that provide the answer. There are many of these stations that are designed specially to help new ham radio operation.

ALE Mailboxes and Bulletin Boards: Internet systems can make contacts and exchange information, and when they do, they act like extended communication interfaces interlinked together just like ham. But ham radio is a little bit different from an internet network in that ham uses tones for transmission and not the usual 1s and 0s as voltages on wire. **ALE** means **Automatic**

Link Establishment meaning that the frequency is being monitored all the time in order to enable other hams to connect easily to the network. Travelers and sailors utilize ham radio where they can't have access to the internet.

Swap Nets: usually established between in-person ham and flea markets. A weekly swap network enables other hams to sell their items or to list and buy things they need. A net control station serves to moderate the process and transaction is usually done over the phone once contacts have been established between two interested contacts.

Joining the ham radio community

Hams rely heavily on their large numbers and simple infrastructure (makeup), and they can easily get their lives together after a serious natural disaster disrupts communications over normal channels. With their large number, seeking for help is easy as members will rally round to get each other out of trouble. Hams engage themselves into local and regional communications that can respond to members' needs in times of need. Hams can organize a **field day** which is an emergency operation that allows ham radio operators to work and help people under emergency situations. Hams are not only useful during emergencies as they are mostly found in other activities ranging from sporting activities, festival activities or many other services that require providing emergency assistance to people. More often than not, there exists a beneficial relationship between stamp collection and ham radio. Hams often exchange postcards (QSLs) that have their call signs, station information and sometimes photos. Hams that deal with stamp collection send these cards across the world. The hams that received the postcards usually acknowledge the postcards by sending stamps of their own. You don't get to know the beauty and the advantage involved when hams share greetings and airmails from across the world until you try it. Hams can try to meet in person (one on one communication) or on the radio. You, as a ham radio operator, must belong to at least one member radio club before you are even issued a license to operate as a ham radio operator (more on licensing later). **Hamfests and conventions** are two other most famous ham gatherings. A **hamfest** is like a market involving ham operators bringing their electronic gadgets to the market to sell. A hamfest can range from a small get-together on a beautiful Saturday morning to large gatherings involving hams from across the world.

Let us examine the differences between two important frequencies on which ham operates on.

Differences between UHF and VHF

VHF meaning Very High Frequency and UHF meaning Ultra High Frequency are two different frequency classes over which the ham radio operates. Getting to know the differences between the VHF and the UHF will help you to make the right decision when you want to identify the best signal for your radio requirement. For reference sake, UHF doesn't travel as fast as the UHF, but UHF may give higher bandwidth. Radio communication occurs over wireless networks (as you can't see any wire connecting you and the receiver). These wireless communications are possible over a wide range of operating frequencies. These frequencies are being regulated – not by individuals – but by governments. For instance, in the United States of America, the radio frequency band is being controlled and assigned by the Federal Communication Commission (FCC). The FCC classified frequencies into four different frequency groups; high band VHF (169-216 MHz), low band VHF (49-108MHz), high band UHF (900-952MHz) and low band UHF (450-806 MHz). The Federal Communication Commission (FCC) in the United States determines who they assign each of these bands to – only after rigorous screening. The main users of these bands are usually radio and TV broadcasters that have obtained license, and other commercial communication services.

Let us examine some of the pros and cons of these two frequencies band

VHF – Very High Frequency

The Frequency modulation (FM) broadcast commonly deploys the Very High Frequency band. Other communication services where the

VHF is most commonly used include; mobile radio systems, marine communications, long-range data communications etc. The Very High Frequency bands include radio waves in the range 30MHz to 300MHz. VHF are not usually disturbed by noises emanating from the atmosphere, electrical equipment issues and a host of other interferences. The VHF band has different bands ranging from low-band to high-band. The lower band VHF of 49MHz are usually featured in cordless phones, transmission of wireless microphones, radio controlled toys and a host of others. The more higher VHF within 54-72 MHz are used in TV channels 2-4 and other wireless systems, especially those used in assistive listening technology. VHF around 76-88 MHz are used in channels 5 and 6. The highest VHF band is the 88-108 MHz bands and are used in Frequency modulation radio broadcast band.

The low-band VHF is not usually advisable for serious applications owing to prolonged radio noise that is observable within these bands. Though, the low-band VHF is still being used till date owing to low cost of equipment to set up. The VHF transmission power is pegged at 50mW unless you want to run an assistive listening technology in the 72-76 MHz range.

High-band VHF is common for serious applications. The 169-172 MHz band, which is the lowest high-band range, has up to 8 different frequencies assigned by the FCC and is mostly consumed by the public and most wireless microphone gadgets. The high-band frequencies are mostly referred to as "travelling frequencies" owing to their wide usage across the USA without any fear of it being interrupted. 50mW power restriction is applied. Low-band VHF ranges are less disrupted than the high-band VHF ranges.

UHF – Ultra High Frequency

The Ultra High Frequency radio waves are especially shorter in length than the Very High Frequency (VHF). The UHF measures around 12 to 24 inches with a consequent reduction in antenna length and radio range. The UHF transmission is easily interfered with- that even something as trivial as building or human body can disrupt the normal transmission. The UHF has larger bandwidth occupation than the VHF which brought about a wider frequency range with a high range of audio signals. The power restriction is not as bounding as the VHF – at least 250mW power is allowed which is more than the 50mW allowable in the VHF.

There is little degree of overlapping between the low-band UHF (450-536 MHz) and the high-band UHF (470-806 MHz). Literally, UHF TV channels 14 through channel 69 and business services run with the UHF frequencies. There is a little level of disturbances in the high-band UHF (above 900MHz) and usually needs antennas measuring between 3 and 4 in inches.

The radio waves being generated by the UHF don't usually go beyond your normal line of sight. Any materials or interferences within your sight (such as high-rise buildings, high trees and any other disturbances) will mostly usually affect the frequency range of the UHF. The UHF transmission has high enough penetration to penetrate through high buildings. This makes indoor reception possible. The UHF is often limited (in usage) in some instances due to its limited line of sight. VHF gives a broader frequency range (for broadcast) which makes it preferable in some industrial operations.

The UHF radio signals and reception are especially useful in most aspects of life ranging from GPS, Wi-Fi, Bluetooth, cell phones, cordless phones, television broadcasting and satellite.

UHF features a much higher frequency which leads to a reduction in wavelength – this is of advantage. The size of the radio wave is a function of the length of transmission which consequently relates with the reception antenna. Essentially, UHF antennas are mostly short and wide.

It is mostly very costly to operate UHF equipment when compared with that of the VHF equipment. The reason is because it is not especially easy to build UHF compatible gadgets, owing to the interferences of high frequency and short wavelength radio waves interacting together.

Building your Ham radio Equipment

The ham equipment are often kept in a compartment called a radio **shack.** For most ham radio operators, the shack literally contains one or two hand-held radios while other ham radio operators use cars to operate on-the-go. Vehicles, mostly cars, can function as a good ham shack as most operators have a special workplace at home for ham. These are some equipment you can literally find in a ham shack;

- **The rig:** Rig describes a transmitter, receiver or, most frequently, a transceiver—which is the combination of a receiver and transmitter in one device (the receiver and the transmitter were separate pieces of instruments in time past). The transceiver allows anyone to send and receive signals, either across cities or halfway around the globe. You can decide to choose and get a handheld, base station or mobile rig depending on the way you want to be carrying out your ham activities.
- **Computer:** Many modern ham operators have at least a computer in their radio

shack. Computers are used to monitor and control many radio functions and activities. Modern computers literally aid digital communication. Some ham radio operators use more than one computer for their jobs depending on job requirement.
- **Microphones, keys, and headphones:** These gadgets are very useful in radio activities to improve audio qualities. The radio shack can contain more of these gadgets depending on the hams preferences and choices.
- **Antennas:** The radio shack can also contain controllers and switches for antennas that are located outside the shack.
- **Cables and feedlines:** Wires and cables are important for radio transmission.

CHAPTER TWO

GETTING MORE FROM HAM

Feedlines and Filter gadgetry in ham

The electrical conditions and abilities inside your feedline can be evaluated by radio and antenna tuners. This electrical condition is measured as the **standing wave ratio (SWR).** The standing wave ratio is voltage ratio and shows you how much of the power that was supplied by the transmitter is actually getting radiated by the antenna. There is a built-in meter showing the feedline standing wave ratio. You can incorporate a stand-alone standing wave ratio sensor (typically referred to as standing wave ratio meter or bridge). The sensor will be able to measure the standing wave ratio when you are working on an antenna. A **power meter** can also be used to measure feedline condition which will help you to evaluate the exact power flowing. The standing wave ratio meter is not as expensive as the power meter; though the power meter is more accurate and precise than the SWR meter. You should know which one you are going for while making your purchasing decision. These meters are usually deployed right at the transmitter output.

Filters

Filters are designed to accept or send back ranges of frequencies. Most filters are planned to accept or reject only specific frequencies. Filters can contain inductors and capacitors, collectively called **discrete components**, or even made up of sections of feedline, referred to as stubs. The following are the categories you would ordinarily come across:
- **Feedline Filters:** You can use a feedline filter to eliminate or reduce undesirable signals from getting in the way of your ham radio. These unwanted signals usually emanate from the antenna, and some other

times the ham radio can also affect the antenna signal. When the signals have already been transmitted, you can use the feedline filter to make sure that the transmitter does not radiate any unwanted signal which can cause interference to other devices. Receiver performance can also get tampered with when the receiver comes in contact with these unwanted signals; you can also use a feedline filter to eliminate any unwanted signal that can compromise the receiver.

- **Receiving Filters:** Receiving filters are usually installed inside your ham radio, and are often made from a quartz crystal or more often from a small tuning-fork-like structure. Receiving filters are able to remove all unwanted signals except a single desirable signal in the receiver. The receiving filter improves the receiver's selectivity making the receiver to be able to receive the single desirable while other signals are still available.

- **Audio Filters:** You can use audio filters which are on the receiver output to give extra filtering ability, rejecting any nearby signals or unwanted noise and disturbances.
- **Notch Filter:** The notch filter serves to eliminate a very narrow range of frequencies, such as a single interfering tone.

Getting your Ham on the Air

The ham radio technology uses more than the equipment to make contacts and pass useful information. The ham radio also makes use of these technologies;

- **Modulation and Demodulation:** Modulation simply refers to the technique involved when you add information to a radio signal in order to be able to transmit a desirable sound wave over the air.

Demodulation, on the other hand, is the process involved when you retrieve information from the signal that had already received such information. There are two special types of modulations involved for ham radio which are the **amplitude modulation (AM) and the frequency modulation (FM).**

- **Modes:** Modes are actually some specific types of modulation. You can pick from several modes during transmission, most of which can include data, video, voice or Morse code.
- **Repeaters:** Repeaters are usually relay stations placed strategically to listen to sound on a certain frequency and then transmit the sound on a different frequency. Repeaters are mostly positioned on towers, high-rise buildings, or top of hills; hence they can allow ham use low-power radio mode to converse or send information over large distances. Repeaters

can be linked by radio or by internet to further improve and aid communication across the world. Repeaters can do what is called a **duplex operation** where they are able to listen and transmit sound at the same time.

- **Satellites:** Some amateur satellites serve as "repeaters in the sky (can listen to sound on a certain frequency and then transmit the sound on a different frequency)," while most other amateur satellites are deployed as orbiting digital boards and e-mail servers.
- **Computer software:** Computers have integrated with almost all facets of lives. Nowadays, most activities are being monitored and controlled with computers because they can transfer and store information at a very fast rate and huge sizes. In just the same way, computers have been integrated with ham radio. In times past, hams were only able to carry

out paperwork while limited with the gadgets inside their sack. But nowadays, computers are now part of the radio gadgets and are able to generate and understand radio signals, sending Morse code and monitoring other essential radio functions.

Ham radio operators have forever been interested in developing radio technologies to aid their efficiency and make their work less tedious. This is one of the main reasons why ham radio is actually a licensed service (you need to get a license to operate one). Hams, today, are in the process of creating a novel mixture of radio and other advanced radio technologies, like the GPS radio location and the internet technologies. For example, the Multimedia group of the American Radio Relay League (AARL) is working tirelessly to incorporate wireless Local Area Network technology in ham radio. The Tucson Amateur Packet Radio (TAPR) has members all over the world who are keen at developing novel means of digital communication.

How natural environment affects radio transmission

The natural environment affects radio transmission and propagation since radio waves use the natural environment as route to space or to their destination (sometimes to another terrestrial station). To make local range contacts, the radio wave moves across the earth surface in a process called **ground wave propagation.** The ground wave propagation can aid communication up to hundred miles. For longer range contact, the radio wave passes through the atmosphere. The upper part of the atmosphere containing high concentration of ions (ionosphere) reflects these waves and sends them back to earth. This often occurs at High Frequency (HF) and Very High Frequency (VHF). Radio wave reflection is termed **sky wave propagation** and the sky wave path can be as extended as 2000 miles (depending on the angle from which the signal was bounced). High Frequency signals normally traverse between the ionosphere and the earth surface multiple times to enable contacts to be possible all over the world. A difference case occurs at Very High Frequency where multiple bouncing is not so common. The ionosphere is not the only contributing environmental effect; as the atmosphere also reflects radio signals and waves. The atmospheric contribution is in the form of **tropospheric propagation** which is common along weather fronts, thermal inversion and major other surface phenomena. The tropospheric propagation is often common at Very High Frequency and at Ultra High Frequency. The aurora and the meteor trail which are part of the VHF and UHF reflecting features also affect radio signals. If the aurora is strong, it will absorb High Frequency (HF) signals, and sends back the Very High Frequency (VHF) and Ultra High Frequency (UHF) signals.

Tip: if your ham is active on these bands, you can point your antennas to the north side to check if the aurora can aid unusual contact. Meteors are usually so hot that the gases start reflecting signals until they cool down.

Joining ham clubs

Finding other hams

One of the beauties of the ham community is on-boarding new members and providing necessary assistance to them in times of distress. New hams usually take advantage of the assistance being rendered to them to make way out of hams and help themselves navigate the ham world. If you are new to ham, there are many ways you can stay in touch and get to know other hams which will be discussed below;

Radio clubs

Local radio clubs are set up to help hams (both new and experienced hams) to stay connected. The first appearance of the radio club dated far back as the olden days when a group of young experimenters came together to build radios – even at the time when the technology was not yet booming. With time, the radio club developed and started growing in number and relevance. As at now, the club ranges from small to various large groups with similar interest.

The following keys are true for most hams you would come across;

_ **Most ham operators belong to at least one club or several clubs at a time:** Most popular and unpopular hams belong to one or more special clubs where they get to meet for chit-chat and to discuss ways of taking the ham community forward.

_ **Most local or regional ham clubs usually engage themselves for in-person meetings:**

Regional clubs are not always as large as international membership.

_ **Specialty clubs are primed with activities:** There exists a large membership base for groups with wider activities ranging from sporting, Ham TV, and other mid to high budget activities.

_ **small chapters may have no reason to conduct in-person meetings:** they can have an over the air meeting, if they want, but transporting themselves to a meeting venue might not be necessary since the membership is not really large.

Ham clubs are great avenues where you can readily get assistance and guidance. As a beginner in ham radio, you will need someone or people who will be willing to devote their time to put you through some basic ham things. Where else can you get such help if not ham club?

Finding and choosing the right ham radio club

To find area-specific ham radio clubs that you can join at no cost, simply navigate to http://www.qrz.com/clubs.html and choose your state to have access to the list of the state's radio club Websites. For example, if you are domiciled in Washington, the Inland Empire VHF Radio Amateurs - Located in Spokane is a good radio club for you. You can also check the website of The American Radio Relay League (ARRL) to have access to their directory of affiliates. You can access the American Radio Relay League's website on http://www.arrl.org/FandES/field/club/clubsearch.phtml. Once you are on the website, simply input your state, city, or zip code to find clubs that are nearer to you. Put emphasis on the general interest clubs as a beginner and search for clubs that can offer guides to you as a new ham radio operator. Focus on the general interest clubs and look for the clubs that offer help to new hams.

If you join a radio club, you have a lot of programs and activities you can benefit from and contribute to. In an instance where you have more than one radio club in your area, you can consider these tips to choose the best radio club for yourself;

_ **You need to give priority to the club whose meeting time and schedule is most convenient for you:** let us say you used to have another business on Mondays – maybe office work or other functions – and club A schedule their meeting time on Monday mornings. You are not advised to choose Club A.

_ **Which of these clubs has activities or programs that best represent your interest?** You can check for the club's monthly review on their website to have access to the list of activities they normally engaged in. Check these activities and be sure they are what you would ordinarily like to do.

_ **is the club comfortable for your kind of personality?**

You can actually attend one or two meetings to have an idea of how the club operates. By doing so, you will be able to have a glimpse and quickly make a decision if it is a club you would like to join. If you cannot attend their meeting, you can browse through their website and newsletter to understand what they do and how they do it.

Improving your club participation

After you have successfully identified a club of your choice, and you have even started attending meetings. Your next task should be finding friends in the club who will be willing to put you through. Follow these tips to do well in your next ham meeting;

_ **Make punctuality a friend:** As a beginner, you cannot trivialize the importance of coming to the meeting very early. This will show your colleagues that you are ready to learn and move the group forward even while you are still new to the organization. It is not a bad thing if you decide to come early to the meeting so that you can help arrange the meeting tables, set the banners and run some errands. You might as well be penalized accordingly if you are in the habit of coming late to the meeting.

_ request for logbooks, if any, where you can sign up as a newcomer or sign in upon arrival.

_ request for a tag that will show your names, call sign (more on this later) and other identification symbols that will allow group members to easily identify you.

_ As a beginner, ensure you do a proper introduction to all the group members – starting with the gentle man sitting next to you.

_ meet with the president, as a first timer, and then introduce yourself to him. If, during the meeting, the president requests you to stand up and identify yourself; kindly oblige and the introduction with greetings followed by your names.

_ get to know the club committees and re-introduce yourself to them. As them if there is any task they routinely assign to new club members.

_ attend group functions and parties!

The tips above don't only work for ham groups, but you can also apply them to every other group that you belong to so that you can get the best out of the group's activities.

Getting involved in ham radio group

As a beginner, you will most likely be wondering how you can familiarize yourself with most of the group activities and get started with the group. The following are some places where you can get involved, though the list is not exhaustive and your group activities might be more depending.

_ **Field days:** This event normally comes up in June every year. The event organizers and planners will most likely need some helping hands. Endeavor to join them and lend a helping hand. This is a good way to get to meet many members of the group and have some chit-chats as you carry on with the task.

_ **Conventions or hamfests:** If your club normally holds a club event, your help is most likely needed. The hamfest is another great opportunity for group members to come together.

_ **Awards and club insignia:** If your club has a group business, you can opt to help them manage sales, take orders, keep records and even make sales during group meetings. If you have another that you are willing to share, don't forget to do so.

_ **Libraries and equipment:** Your club might have a library where members get to read and borrow books. You can volunteer to manage the library or keep track of the book being borrowed by other members.

_ **Club stations:** If your club is good enough to possess its own radio shack or repeater station, you can offer to help with some maintenance work — such as working on antennas, tuning and testing radio or just some routine cleaning tasks

you can team up with the station manager to get familiar with the equipment being used very quickly. You don't need to have every technical knowledge of these pieces of equipment; only your willingness to help and learn is needed.

If you have the wherewith to write or create Web sites, you should not hesitate to help with the club website or newsletter. It is possible that they have many things you can actually help with.

Here are a few common club activities you can get acquainted with;

_ **Public service:** This activity normally involves providing local communication

during an event involving sporting or public gatherings, such as a festival or parade. These events are excellent ways for you to show and improve your communication skills.

- **Contests and challenges:** Operating events are nice fun and most clubs Participate in many on-the-air contests as a team or club. Sometimes, you can have a club challenging another club over which club will be a winner and a loser.
- **Work parties:** The primary purpose of a club is to assist other members. Building a radio station with other group members might be a good means of getting to know each other's skill and maximizing it for potential benefit.
- **Construction projects:** Clubs may sometimes come together and contribute funds to sponsor projects which are meant to be done individually. The club gets to save money by building their own equipment and this enables group unity.

Supporting your club by actively engaging in activities and committees is very important.

For one thing, you can acknowledge the help you get from the other Members. By being an active member within the club, you can strengthen the club, your friendships with hams, and the ham operation in general.

The **American Radio Relay League** *(ARRL)*

The **American Radio Relay League** (ARRL) is the oldest functioning amateur radio organization in the world. It was founded before World War I. The **American Radio Relay League** extends services to ham radio operators all over the world and plays an essential role in actively representing the ham radio mission to the public and governments in general. You cannot ham radio surviving and still functioning for more than a century without a functioning leadership like the **American Radio Relay League**.

What can ARRL do for Hams

The most obvious benefit of the **American Radio Relay League** *(*ARRL) membership is that you get to receive *QST* magazine once in every month. The *QST* is the largest and oldest ham radio magazine in the world. The magazine usually includes feature articles on both technical and operating topics; insights on regulatory parameters affecting the ham radio as a hobby; the results of various competitions sponsored by the ARRL; and some other columns that cover a wide array of topics. All of the biggest ham radio equipment makers in the world advertise their products inside the *QST*. There are other interesting ham radio magazines; only that *QST* is the most widely read and of great relevance.

Along with the *QST* magazine, the **American Radio Relay League** also maintains a functioning and substantial Web site at www.arrl.org. You can explore the following features on their
Web site:
- News and general interest stories that are normally shared every day
- The Technical Information Service- which is an extensive reference service-featuring technical document searches having numerous articles searchable online

- An active ham-radio swap and shop which is available throughout the day.
- Many free email bulletins and featured newsletters

The **American Radio Relay League** Field Organization anchors the activities of teeming volunteers. These volunteers are grouped into eighty different sections in fifteen divisions.

The **American Radio Relay League** Field Organization also features the world's most comprehensive nongovernmental radio channel - the *National Traffic System* (NTS). The NTS is charged with the responsibility of actively responding to various emergencies and is available on a daily basis, sending radio messages, or regulating traffic, all over the world.

Together with administrative and organizational responsibility, the **American Radio Relay League** is as well the biggest single sponsor of most operating activities for ham radio operators.

Special purpose Ham organizations and clubs

Ham radio is huge, deep, and wide. A **specialty club or specialty organization** deals with one part of ham radio that focuses certain technologies or certain types of operation modalities. To find some specialty clubs that are close to you, navigate the Internet by inputting your area of interest and the search word **club**. For instance, when you enter MHRC - Mount Abu International **Ham Radio Club** at www.google.com will give you the idea of the clubs that you are looking for. Some clubs based their operation on particular operating interests, like qualifying for awards or operating on a single band. An example of the latter is the 10-10 International Club (www.ten-ten.org/), which exists for ham operators that want the 10 meter band, a favorite of low-power and mobile stations. The 10-10 Club funds many contests every year and gives some set of awards for contacting its members. The Six

Meter International Radio Klub (SMIRK) also promotes function on the 6 meter band with its special and unusual methods of signal propagation. Another example of a specialty club is the **contest club.** The **contest clubs** members love engaging in competitive on-the-air activities referred to as radio sports, or simply contests. The clubs challenge each other, fund awards, award plaques, and also encourage its members to build up stations and methods to rise to top contest operators. The Yankee Clipper Contest Club, one of the oldest clubs in the United States, has a list of contest clubs in the United States which you can found at www.yccc.org/ Links/Contest_clubs.htm. The broad list of clubs that compete in the ARRL competition can be accessed at www.arrl.org/contests/club-list.html. To have access to the list of contest clubs in the world, go to www.ac6v.com/clubs.htm#DX. The long-distance communication specialists, also called the DXers, are not as competitive as other contest operators – but specialize in long-distance contact. Long-distance communication specialists (DXers) form club that are willing to share their operation experiences.

Handi-Hams

This is ham for disable people. Disability is never inability, but if you feel you are disable in any way and you love ham as a hobby – then there is a place for you. Ham radio leverages excellent communication avenues to individuals who otherwise find themselves reduced (somewhat) by physical limitations. Handi-Hams, which was established in 1967, is a specialty club charged with the responsibility of providing equipment and opportunities to make ham radio accessible to people with one disability or another. Handi-Hams not only aids hams that have disabilities reach out to other parts of the world, but as well helps members to team up with other ham members and to diverse helpful services. The Handi-Hams website is accessible and provides links to an extensive group of resources. You can access handi-ham at www.handihams.org.

AMSAT

AMSAT (code for Amateur Radio Satellite) is an international organization charged with the responsibility to manage and coordinate launching of satellites and also oversees the construction of its own satellites. These Amateur Radio satellites deploy amateur radio frequency allocations to aid communication between ham radio stations.

Many amateur satellites have an **OSCAR** tag. The **OSCAR** is an acronym standing for **Orbiting Satellite Carrying Amateur Radio**. The tag is given by the **Orbiting Satellite Carrying Amateur Radio** (AMSAT), which is the organization promoting the development and launch of all amateur radio satellites.

Because of the widespread importance of this tag, amateur radio satellites are sometimes called OSCARs.

The Amateur Radio Satellites can be used without charge once you are licensed ham radio operators for voice (FM, SSB) and data (AX.25, packet radio, APRS) communications. As at the moment, there are more than 18 fully active amateur radio satellites in orbit. They may be tagged to function as linear transponders, as receivers, and as just a store and forward digital relays.

Amateur radio satellites have aided the advancement of the science of satellite communications. Impact includes the building and launching of the first satellite voice transponder (OSCAR 3) and the building of very advanced digital "store- and-forward" messaging transponder methods.

The Amateur Radio Satellite has an active community which is very proactive in creating satellites and in exploring launch opportunities.

New satellites are being launched everyday which make it necessary for the list to be updated on the AMSAT website every time. This is the responsibility of AMSAT.

Hamfests and Conventions

Hamfests are one of the best and most exciting activities in ham radio. Hamfests are ham radio flea markets where the primary sellers are people displaying gadgets from a folding table. Most hamfests are very small and don't last for long – just some few hours, while some other hamfests curate many buyers and sellers lasting several days. Hamfests activities can either be indoor or outdoor activities. A ham radio convention is more packed with activities than a hamfest. It may sometimes include licensing test sessions, speakers, seminars and commercial sellers and buyers. Conventions sometimes feature a swap meet together with the rest of the functions. Conventions often feature a central theme, such as QRP or low power operation, emergency operation or digital radio transmission.

Finding hamfests

The best website where you can access Hamfests in the United States of America is the **American Radio Relay League** (ARRL) website at www.arrl.org. Navigate to the top part of the home page once you have entered the website. At the top part, you will see a hamfest link which will take you directly to the hamfest and convention page. You will see an ARRL section where you can search for events by states or divisions. After you have identified your hamfest, set an alarm for the specific time you want to be attending the meeting. Be aware that most hamfest events are Saturdays-only events.

Don't forget to take the following materials with you while going for hamfest

_ **An admission ticket:** Some hamfests require you to display a ticket at the entrance before you are allowed entrance to the event. You can get the ticket at the gate upon arrival or you can order straight from the website.

_ **Money:** You may decide to buy something during the flea market, which is why you are advised to take your money with you. Some Hamfests don't accept cards.

_ **Take a bag with you to carry your purchases:** You will need to contain your purchases. Electronic gadgets can be kept in a sack or bag.

_ **A hand-held or mobile rig:** Some hamfests have a talk-in frequency,
which can be a VHF or UHF repeater. The mobile rig will come in handy if you don't know how to navigate the area; enabling you to get direction very easily.

_ **Water and food:** Though there will most likely be a food stand where you can get some quick food, you can't particularly count on this.

It is advisable you come with your own food and a bottle containing safe water for drinking.

Buying at hamfests

Once you have finally made it into the flea market, you can start making some purchases. But you don't just start buying without having some basic information about the items you are trying to buy. Consider the following tips to help you out:

_ try to team up with an experienced ham that can help you navigate the ham market.

_ You can negotiate your prices and get a better deal.

_ You can get to buy more accessories for your radio at a price lower than what the manufacturer will sell them.

_ If a vendor doesn't allow you to inspect the gear you want to buy very well, it is a good sign you should not buy from him.

_ Don't buy a burnt or overheated gadget. This is why initial inspection is necessary before you invest your money.

_ If you don't know what a gadget is used for, be ready to inquire from the vendors. They are always ready to tell you everything you need to know about the piece. You might not necessarily buy the gadget in the long run, but you have just made a new friend.

_ there are many auction websites for ham where you can get to see the price and condition of the gadgets you want. Some examples of such websites are www. eham.net, www.qrz.com, and www.arrl.org/RadiosOnline/.

Finding conventions

You can get to know about any scheduled convention just about the same way you know of hamfest. But one important thing to note is that conventions are usually bigger and tend to consume a lot of money than the hamfest. Conventions are normally done in an hotel and it is usually advertised in most ham magazines. The two biggest ham radio conventions are the Hamvention, held in Dayton, Ohio, in mid-May, and the International Amateurfunk-Ausstellung, which is normally held in Friedrichshafen, Germany, in late June. Dayton convention mostly features 25,000 or more while the Friedrichshafen convention is about the same population also. The ARRL National and Division Conventions (which you can access on the ARRL Web site) are usually conducted in every region of the United States. The Radio Amateurs of Canada also has their national convention which usually takes place every year. These conventions typically get attendees ranging from a few hundred to a few thousand. The conventions are meant to be family friendly. The conventions as well provide a place for specialty groups to conduct conferences within the main event. These smaller conferences are places you can normally find extensive programs on QRP, direction finding, county hunting, wireless networking on ham bands, etc. Some conventions and conferences discuss one of ham radio's many aspects, such as VHF, UHF, DX-ing, or digital technology. If you are an avid fan of a particular event or activity, attending a weekend convention is a good way of meeting other ham radio operators with similar interest like yours.

Tips for ham radio operators

You can check the following useful tips to get the most out of ham;

Listen to everything: Most people don't know the benefit involved in listening more and saying less. Every moment you used to listen to other professional hams is every minute you used learning new things. Learning is continuous, even if you think you have understood everything, you will still find some people that know more than you. Be teachable and adaptable, only then you can see how far you can go. This not only applies in ham but also other professional endeavors.

Follow the protocol: You can't disregard the importance of laid-down protocols if you are planning to get the best out of ham. You should use appropriate terms while addressing other hams and send information using the right form and formats. When you send a call to another ham station, adopt the "Gift Tag Order – To then from." Begin with that particular station's call sign to notify that operator, and then send your call once or two times as necessary.

Practice to make perfect: You should not underrate the importance of consistent practice. Your work as hams require you to understand some basic terms of communication, familiarize yourself with these terms through practicing. Keep yourself in tune with the radio world by doing regular radio exercise.

Pay attention to detail: Small details are as important as bigger ones. Always try to pay close attention to everything as hams.

CHAPTER THREE

LICENSING THE HAM

Ham radio is not like other types of radios out there because you cannot get to use ham if you have not obtained the license from relevant quarters or government agencies. A system without order is like a porous system where people get to do anything they want without anyone challenging them. To avoid unnecessary porosity, the Federal Communications Commission (FCC) has mandated that anyone that wants to operate a ham station must go through the licensing process. It is not only enough to get the required license, it is also important you keep to the FCC's guidelines if you want to keep and maintain your license. Failure to obey basic regulations can lead to termination of your license or other penalties. A ham license, essentially, is an authority issued to you (as a hams) to use some portion of the frequencies under the FCC for the purpose of transmitting radio information.

Note: you will have to take an exam or test before you will be granted a license to operate (more on this later).

This chapter will take you through all the steps involved to get your ham license starting from the basic requirements to the advanced one.

Frequency assignment

The International Telecommunication Union (ITU), established in 1932, serves to maintain orderliness within the radio community. The International Telecommunication Union is a non-governmental organization (NGO) is a platform for formulating the rules that guide the usage and assignment of radio spectrums. The ITU divides the radio spectrum into ranges and then allocates to some specific users. The ITU allocates spectrums by categorizing the world into three categories based on continents. These categories are referred to as regions. Within each region, each type of radio service — military, amateur, government, and commercial usage — is assigned a specific spectrum of the available frequencies. Luckily for amateur radio operators, most of their spectrum assignments are about the same thing in almost all the three regions. Ham radio operators have little allocations at various places in the radio spectrum and permission to those frequencies, even, depends on the category of license class you possess as hams. The more the class licenses that you possess, the more frequencies you can have to yourself.

Licensing classes

When you take more exams, you will be given more frequencies and privileges to operate ham. There is a certain test level called **element** which when you pass you receive a credit for it (just like a thumb up). After you have taken a license exam, your license is valid for a period of ten (10) years and you will be able to renew such license without any qualifying test. The American Radio Relay League (ARRL) and many other organizations have study guides and manuals that you can use to get past the test stage. Some of these manuals and guides are available through your local library. Depending on your location in the United States, there are many local libraries and you can shop around at your leisure. You only have to make sure the test materials you are studying are the current material because the test actually changes from period to period. When you have taken your time to go through the test materials, you can be assured of success on the exam day.

In the United States, the below are the types of licenses being granted to hams which include; Technician, General, and Amateur Extra.

The Technician License

This is like an entry-level license class if you are just getting started as a beginner in ham. You will be required to pass a qualifying exam of about 35 questions. You are expected to get 26 or more questions correctly to pass the test. This exam tests your ability on radio theory, radio regulations and operating procedures and practices. With this license, you will be able to get an access to amateur radio frequency above thirty (30) megahertz. This frequency (30MHz) gives whoever has been assigned an opportunity to communicate with other hams and people within the local community and quite often within North America. It also gives you some

sort of limited privileges on the short wave (HF) bands that you can use for international communications. If you can also pass the 5-word-per-minute (wpm) Morse code exam, you will also get some transmitting access in parts of four of the normal shortwave HF bands. The four traditional short wave bands you can have access to include;
- 80 meters band (frequency privilege around 3.675–3.725 MHz)
- 40 meters band (frequency privilege around 7.100–7.150 MHz)
- 15 meters band (frequency privilege around 21.100–21.200 MHz)
- 10 meters band (frequency privilege around 28.100–21.300 MHz for Morse code, RTTY, and data and 28.300–28.500 for Voice)

You can write your Morse code exam at the exact test session as the technical license written exam or at another date if you desire. Morse code is needed for amateur operation that is below 30 MHz due to an international treaty introduced some years ago. In those years, a huge number of commercial and military radio traffic — telegrams, news, ship-to-ship, and ship-to-shore communication – was done with the Morse code. Morse code was seen as a standard radio skill in those years. But recent modification has seen this treaty not reckoned with anymore, and Morse code is losing its stand in amateur radio licensing. Nonetheless, Morse code still contributes to a large part of amateur procedures and operations; ranging from normal messages, ragchewing, emergency operations and various contests.

The General License

The General class license gives some privileges in terms of operations on all Ham Radio bands and virtually 99% of, if not all, operating modes. The General License gives you permission to international communications. Just like the technician license discussed earlier, the general will require hams to pass a 35 questions test. If you want to have the General license, you must have taken and passed the Technical license test. The General class test, although more detailed, has the same topics just like the technician test. When you get a General class license, you have achieved a huge milestone.

All of the main frequencies on the High Frequency bands are also applicable to General class licensees.

The Amateur Extra License

This type of license class or category gives you all of the available United States Amateur Radio operating permissions on all frequency bands and all operating modes. If you are planning to get this license, you need to be reminded that it might not be as easy as the first two licenses. It involves taking and passing a 50 questions test. To get this license, you must have written and passed the technician license exam and the general license exam. The amateur extra class exists because the general class might not necessarily have all the licenses. The lower part of many High Frequency bands are reserved for the amateur extra class licensees alone. These lower parts of the frequency bands are where you find the ham experts.

You are required to pass 37 questions out of the 50 questions before you can be successful for this license class.

Grandfathered license classes

The past few years have seen amateur radio licensing rules metamorphosed in order to bring down the number of available license classes. If you have a license in any of the license classes that have been deleted, you may be able to renew your license. At the moment, you will not be able to get a license in those license classes that have been expunged. This is why the grandfathered license classes exist. There are three categories of the grandfathered license classes which include;

- **Novice:** This type of grandfathered license class was introduced in the year 1951. To qualify for this license, you will write a simple 20 questions exam and a 5 word-per-minute Morse code exam. The test will be administered to you by people in the General license category. At inception, the novice license was meant to be valid for one year after which you will be asked to upgrade. At the moment, though, the license now has a validity period of 10 years and you can renew it. Hams with this

license are only permitted for the 3.5, 7, 21 and 28 megahertz bands.
- **Technician-Plus:** The technician-plus class is not exactly different from the current technician license, where you are required to take and pass a 5 word-per-minutes Morse code test.
- **Advanced Class:** This exam is somehow hard – just in between the general class and the amateur extra class. They get frequency permission more than licensees under the General class but not as much as the amateur extra class.

Taking the license test

If you want to study effectively and avoid studying unnecessary materials, then you need to understand how the test is compiled. You will take a multiple choice test for your license class exams (irrespective of the type of license class). There is no pictorial or essay question there. The test for each license is usually called **element.** The Morse code test (Element 1). The written test for the technical license is called Element 2. The General test is Element 3, while the amateur extra is the Element 4 will gauge your capacity to receive Morse code.

The test includes four essential areas:
- **Rules & Regulations:** This covers essential rules of the things that you have to understand in order to legally operate ham.
- **Operating:** This includes some useful procedures and conventions that hams follow when they are on the air
- **Basic Electronics:** This covers basic concepts on radio waves and electronics featuring some simple calculations.
- **RF Safety:** This contains questions guiding the installation and operation of transmitters and antennas.

The test has a set number of questions from all the four areas; these questions are chosen randomly from the four areas. The Technician and General tests contain 35 questions, while the Amateur Extra test contains 50 questions.

Finding the right material for study

Once you are ready to get started on the path to taking your test, there are many study materials you can lay your hands on. These study materials range from books, videos, classes, software and online materials. A good example of online materials, though not for the purpose of taking your test, that you can lay your hands on is this eBook you are currently reading. Your study materials can as well be in the form of an eBook that you can get to read at your leisure.

Before purchasing your study materials, kindly note that the test questions and

Ham regulations change occasionally, but sometimes they don't.

Be sure to confirm that the study materials you are getting feature the latest updates. You can consult the ARRL website www.remote.arrl.org/arrlvec/pools.html to get access to the date of the question pools (materials) that are currently available.

Finding the right licensing classes

You can find the right classes for yourself by:

_ **Asking other members at your radio club:** Your club might organize classes which will be sponsored by the club. Take the advantage of such classes where you can learn the most beneficial things for your exams.

_ **Looking for upcoming tests to be conducted in your area:** You can get to find out more about upcoming tests from the American relay radio league (ARRL) at www.arrl.org/arrlvec/examsearch.phtml. You can contact the liaison officer to get more information about licensing tests.

_ **Asking at a nearby ham radio or electronics store:** In the United States, it is quite easy for you to walk into a nearby ham store within your area to ask more about ham tests. You can see a bulletin or website lists around these stores containing some upcoming classes.

Stores that market electronic supplies to people, such as Radio-Shack, may also know some information about ham classes. In addition, you can carry out a Web search trying to look for **ham radio class** or **radio licensing class** very close to your city or state.

Other avenues for you to find classes range from local disaster-preparedness organizations, Schools and colleges that usually provide space for classes, and some public safety security forces such as the police and fire departments. One time saving skill you can learn here is that you can actually take a weekend for these classes, and then take your test the next Sunday and get your call sign having passed the required exam. This is a big time saving skill, especially if you used to be busy during the weekdays. The licensing test can be easy and can be difficult. Some people crammed their way out of it overnight, wrote the test the second day and then got their new call sign in the list inside the database of the Federal Communications Commission (FCC). More often than not, cramming can make you forget most things you have studied just after the exams. But people do not get to really care; after all they have passed the test and gotten the call sign already.

The following are the right material guides for your studies;

Books, software, and videos

If you are studying for a technician license, the best available guide for you is the American relay radio league's book titled "**Now you are talking.**" This book explores just beyond providing you with the questions from the test's pool. The book explores the why and how you have to take the exam and pass. You can get this ARRL's book from www.arrl.org and also from some retail platforms; whether online or offline.

There is a popular hams called Gordon West WB6NOA who has written more licensing instructions for all the licensing tests you might want to take. The Gordon's guide actually features the question and answer format. These guides are meant for you if you want to take and pass the exam quickly, and don't necessarily give you some basic backgrounds just like the ARRL's guide. If you are a video person (loves learning through videos), the best packages for you are the ARRL's Technician class and general class series of tapes. These packages have been prepared professionally for you to learn easily without stress.

If you are a software person; just like me – your guide is the Ham University CDs. These CDs feature a technician licensing course together with a Morse code disc.

Online

There are many online resources for you to get started with, though they might not necessarily be as elaborate as the eBooks and videos. Nonetheless, you cannot trivialize the importance of readily available online information. With the right gadgets, you can source information as much as you can very easily. You can practice the online exams as many times as you want. When you practice before the D-day, you will get used to the test format and this takes away the anxiety. The online scorer will score your test and will even advise you on the test areas that seem to be your weakness. You can consult these websites to have a glimpse; www.aa9pw.com/radio/, www.qrz.com, and www.eham.net.

Finding a Tutor

While you can study for your ham test on your own and still do very well, you cannot, on the other hand, ignore the importance of a good mentor that can put you through the basics. It is possible you get boxed on a very crucial aspect and just need someone to put you through.

As in other similar situations, the right way to troubleshoot a problem is to seek the guidance of someone who is considered to be more experienced than you are to put you through. The person that is ready to put you through (mentor) is called an **Elmer.** Elmer is not a

name of anyone, to be specific, but just someone who can basically put you through the basics. You can find the right tutor in any of these places;
- **Ham radio licensing class:** you can get to ask questions in the class.
- **Radio clubs:** Radio clubs might find the right class for you.
- **Online:** This is a good place for asking questions and sharing ideas, especially if it is the kind of question you don't want to ask in the presence of others. An example of an online chat room can be found at www.qsl.net/n5sdd/.

After you have succeeded getting your license, don't forget to teach others about your experiences and what you have learnt in the process. You don't know how much it might mean to them.

The Volunteer Licensing System

The entire processing of license application is not alone handled by the Federal Communications Commission (FCC) because the body is not the one that is currently supervising and administering the licensing examinations. In the United States, the ham radio license examinations are administered by volunteer examiners who have been certified and found worthy by a coordinating organization: the Volunteer Examiner Coordinator (VEC). The quality and the profoundness of the licensing examination process is not something to be toyed with -quite high. The flexibility, in terms of volunteer supervision, provided by an all-volunteer body actually goes a long way to make the test taking process a walk in the park process. These groups of volunteers handle all of the paperwork involved and even go the extra mile to file the results with the Federal Communications Commission (FCC). Nonetheless, the real license is still within the power of the Federal Communications Commission (FCC). In the past, tests were usually taken at the "local" FCC's office, which could be miles away from your residence. Nowadays, the tests are often administered at locations not too far from a ham club, schools, or sometimes at home.

Volunteer Examiner Coordinator (VEC)

The Volunteer Examiner Coordinator (VEC) is the body that voluntarily takes the responsibility to coordinate the volunteer examiners who supervise the test sessions. The body also processes all of the paperwork required by the Federal Communications Commission (FCC) to grant a ham license. The Volunteer Examiner Coordinator (VEC) that has the highest volunteering persons serving as examiners is the American Radio Relay League-VEC (ARRL-VEC). Aside from the American Radio Relay League-VEC (ARRL-VEC), there are about thirteen (13) other VECs within the United States of America. You can navigate a Volunteer Examiner Coordinator very close to you at www.wireless.fcc.gov/services/amateur/licensing/vecs.html. Some Volunteer Examiner Coordinators (VECs), like the ARRL-VEC and W5YI-VEC, work nationwide, while a lot of other VECs operate within a single region. The Volunteer Examiner Coordinator (VEC) is the one that is responsible for preparing and administering the license test and other necessary materials. The VEC collects the test results, solves all differences, and then collates all of the data with the FCC electronically. This process gives you a ham license and callsign much faster than when the process is handled by the FCC. New licensees usually get to know their call signs just within seven to ten business days from the time of test taking.

Finding a Test Session

It is easier now, than ever before, to find a suitable test session and venue right on your computer and at the comfort of your home. You can search the Federal Communications Commission (FCC) Web site's list of bodies that serve as Volunteer Examiner Coordinators (VECs) in the different part of the United States at www.wireless.fcc.gov/services/amateur/ licensing/vecs.html#vecs.
If you have seen a Volunteer Examiner Coordinator (VEC) in the list that is close to you, you can begin by contacting the body. Many of the Volunteer Examiner Coordinators (VECs) that you are likely to come across have a Web site and every member has an e-mail contact. You can send an email to introduce

yourself by saying, "Hi, I am....(Names) and I would like to take the general license exam... (Name of exam). Kindly forward me the list of test venues and dates. I stay in Washington, United States...." If you visit the website above and still couldn't find a suitable exam time or a VEC, you can try the following Volunteer Examiner Coordinators organization;

_ **ARRL VEC exams:** The ARRL VEC works nationwide and you can look for exams based on your current zip code. Just go to at www.arrl.org/arrlvec/ examsearch.phtml.

_ **W5YI VEC exams:** Like the American Radio Relay League, the W5YI VEC (which was initiated by Fred Maia W5YI) also works nationwide. You can go to www.w5yi.org/vol-exam.htm. to access the list of certified examiners.

_ **W4VEC VEC exams:** The W4VEC (the call sign of the Volunteer Examiners
Club of America) works within the Midwest and southern American states and gives you a
list of dates and venues which is accessible at www.w4vec.com/ar.html.
If you are still finding it difficult to find a convenient exam venue and date for yourself,

you can communicate with the VEC body by looking for their contact on the FCC page www.wireless.fcc.gov/services/amateur/licensing/vecs.html to request for help. They will be willing to help you and assign you a suitable test venue and date.

Signing Up for a Test

Once you have successfully found your preferred test session, you need to address the test hosts to intimate them that you want to attend their session and the precise test element you want to write.

Public and Private Exams

Most license exam sessions are available to the public and are usually taken at schools, churches, or other gathering places. Most of the test sessions are one where you can just walk in without prior appointment, pay the test fee and then sit down for your test. But sometimes, it is always good to announce first before you show up for such a test as you cannot always judge what the current situation might dictate.

Public test sessions are very profound ways to take a test. You can find exams being taken during most Hamfests and conventions. The FCC mandated that test takers should not pay any attendance fee at the test venues, but you may be made to pay for a particular entrance fee as a solidarity fee.

Private exams are also possible as you can sit for exams within your local community or even in your friend's sitting room.

Understanding Call Signs

When you finally get your license after much drilling and tests, your license will feature one very important thing which is your **call sign**. The **call sign** is also called **call to hams.** The call sign is your identity on the air. Once you have been granted a license by the Federal Communications Commission, you will automatically get your call sign along with it. Your call sign will become your unique identifier in the air. Let us say your call sign is NØZA, your colleague will be addressing you with this sign. When you want to pronounce the call sign, you will pronounce each letter and number separately and not as a whole. The symbol Ø is used in place of zero (0) in call sign. Ham radio call signs all over the world are devised from two parts: the prefix and the suffix. The suffix of the call sign, when joined with the prefix, uniquely identifies you. Each call sign is unique. A lot of call signs might consist of NØ and ZA, but just one call sign is NØZA.

The prefix is made up of one or two letters and a numeral. For example, the prefix in the call sign above is NØ. The call sign will also reveal the country that gives you the license and may also show where you live within that particular country. For the United States call signs, the numeral part reveals the call district of your license when they issued it.

The Suffixes always consist of one to three letters without any punctuation character - just A to Z and Ø to 9.) The suffix in the call sign above is ZA. The ITU gives each country a block of prefix character that enables the government to designate licenses in all of the country's radio services. Licensees in the United States (not only hams) all possess callsigns that start with the letters A, K, N, or W. Even broadcast stations in the US have call signs like KGO or WLS. Most call signs in Canada start with VE. You can go to www.ac6v.com/prefixes. htm#PRI. to have access to the full list of all ham radio prefix designations. Your class license will be shown in your call sign. Essentially, the higher your license class is, the shorter and more unique your chosen call sign will be. Your call sign is a proof that you have dutifully taken and passed the licensing exam and you now have the permission to build and operate a station. This is something worth celebrating.

GETTING YOUR LICENSE

Completing your paperwork

It is not enough to write the exam as there are still lots of paper works to do before you finally get to collect your license. The first thing you do as soon as you have completed your licensing test is to fill two forms which include;

- You will be required to complete the Certificate of Successful Completion of Examination (CSCE). You will have your own copy of the completed form and this will serve as the record of your own test credit and you can show it as proof as proof of the exams you have passed if you need to upgrade your license at a later time. Keep a copy of your Certificate of Successful Completion of Examination (CSCE) properly till the Federal Communications Commission (FCC) issues you a new license or save the change to reflect in their database. This form can serve as a record of your achievement.
- The second form you will be required to fill out is called the NCVEC Form 605. The NCVEC Form 605 enables the Federal Communications Commission to process your new license. Anytime you are issued a new license, change to a higher class, renew your license, name or address change, or you are just picking up a new call sign, this form will come handy. You can also change your name, address, or call sign by submitting such a request to the

Federal Communications Commission through mail or online platforms. The volunteer examiner handling the session will send your Certificate of Successful Completion of Examination (CSCE) and the NCVEC Form 605 to the VEC body for certification. You are expected to wait for like seven to ten business days before the FCC updates your record in their database.

Be reminded that you are expected to keep your updated mailing address on record with the Federal Communications Commission (FCC). This is necessary if you change your address or mailing address. The license assigned to you by the FCC is only valid for a period of ten years after which you will be required to file for another copy of the NCVEC Form 605.

Finding Your New Call Sign

As soon as you have completed the required test and paperwork, you can be on the look-out for your new call sign to come up on the FCC's database. The popular question from ham beginners is usually if they can get on air once their call sign appears on the FCC database. The good news is Yes! You can actually get started on air once you have gotten your call sign even if you don't have the paper license with you yet. The FCC operates an online licensee information system named Universal Licensing System (ULS). All licenses have Federal Registration Number (FRN) which serves as their unique identification within the Federal Communications Commission.

You can follow the steps below to properly find your call sign on the FCC web;

1. Navigate to your favorite web browser to access www.wireless2.fcc.gov/UlsApp/ UlsSearch/searchLicense.jsp.

2. Tap on the Amateur link in the Service-Specific column located at the middle
of the web page to have access to the search form

3. Input your last name inside the Name box and your zip code inside the Zip Code section in the Licensee section of the form, and then tap the Submit button. Wait a while for the information to come up.

4. Browse through the results displayed. If the results displayed extend more than a single page, tap the Query Download tab above your search result to see the entire result.

Most of the data you enter during your license application is a part of the Federal Communications Commission's database and the data is available to the public along with your FRN, if you have registered for one. By submitting your application, you have agreed to identify yourself as a ham (licensed) and tell where your ham station is situated (your ham station is not a physical station). The Federal Communications Commission will not show some of your information, such as email, phone number or your social security number, to the public.

Rather than searching the FCC website, there are many websites that you can search to have access to the same database on the FCC website. These websites include;

- **QRZ.com (www.qrz.com):** This is perhaps the best-known ham radio Web site. All that you need to do here is just enter your call sign inside the "Get call sign" box which you can find at the top left side of the web page. In a situation where you have forgotten the call sign, you can actually search by name by entering the name inside the "Name search link" located in the "Get help" section at the left. The information you obtained from this website is straight from the Federal Communications Commission database. This website can as well allow you to search for other hams that are not within the United States provided that their information is available online.

- **American Radio Relay League, ARRL (www.arrl.org/fcc/fcclook. php3):** Enter your last name, location's zip code and then hit the submit query tab to search.

Registering with the FCC online
The Federal Communications Commission has helped a lot by creating the Universal Licensing System (ULS), thus bringing the registration online. This is a lot more convenient for hams as it is faster and easy. With the ULS, you can renew your license, change your address, and perform some other basic things. If you want to

start using the online system, you must be registered in the CORES, which is the Commission Registration System. You can even register if the FCC is yet to approve your license. To get started with CORES, you will need to enter your Tax Identification Number (TIN). The TIN will serve as your Social Security Number. Once you supplied your TIN, you will get the Federal Registration Number (FRN) which will serve as your unique identification with the Federal Communications Commission. Each call sign is normally linked with a license ID which is the identification of the person applying for the license. Once you get your FRN, you will be able to link the FRN with a call sign you have. The following steps will get you registered with the ULS/CORES;

1. **Navigate to <u>www.wireless.fcc.gov/uls/</u>. To have access to the ULS web page.**
2. **Select the CORES/Call Sign menu.**
3. **Choose any of these options and tap the continue link;**

• **Register Now:** Choose this option if you have not registered with the ULS in time past.

- **Update Registration Information:** Select this If you need to change any information.
  - **Update Call Sign/ASR Information:** Select this if you need to either add or remove your call sign that has been associated with your FRN

4. Choose the "An Individual" option and select the precise location of your contact address, then tap the Continue button.
5. Input your name and your address.
6. Input your Social Security Number if you are actually registering for the first time.
7. Enter a password. Don't use your call sign as a password as anyone can easily guess it.
8. Enter any personal information inside the hint box.
9. Tap the Submit button to finish.

10. Correct any errors.
You will be greeted with a window showing your license ID, password, and personal identifier. You can print this page for reference.

CHAPTER FOUR

USING HAM IN EMERGENCY

One of the primal functions of ham radio operators is seen mostly during emergency cases. Emergency can occur anytime, so it is advisable that you start preparing for your first emergency call as hams as soon as you get your license. Emergency communication, also referred to as **emcomm** in the radio profession, is any radio communication between two or more persons with the intention of alleviating a severe threat of injuries or property damage. As a ham radio operator during an emergency, you will be in the position to report any car accident or disaster. The largest nationwide ham radio emergency communication in the United States is the Amateur Radio Emergency Service (ARES). To have access to the complete understanding of what ARES does, you can explore the Public Service Communications Manual at www.arrl.org/FandES/field/pscm. Aside from the Amateur Radio Emergency Service (ARES), there is also the Radio Amateur Civil Emergency Service (RACES), which is an emergency ham organization that is being managed by the Federal Emergency Management System (FEMA). The Radio Amateur Civil Emergency Service (RACES), just like ARES, is a national emergency communications body charged with the responsibility of providing communication assistance to both the public agencies and the private agencies especially during a disaster. Both the ARES and the RACES are opened to participation from interested hams that are willing to join and contribute their own quota. If you are interested in joining the Amateur Radio Emergency Service (ARES) to help with any of their activities, you can consider going for the ARRL field organization role. The ARES volunteers have the following duties;

- **Assistant Section Manager (ASM):** As a new ham, you can always opt to assist the section manager. Tasks differ according to the activities of the section you found yourself, but sampling and analyzing volunteer reports or working on and checking into local and regional networks are likely duties you can be helping with. If there is any special task available, you

might be required to solve it on behalf of your section manager.
- **Official Emergency Stations (OES):** Duties dependent on your local emergency coordinators. The Official Emergency Stations (OES) appointments are meant for those stations that are passionate about emergency works. Their works are often that of logistics, operations or administrations.

- **Public Information Officer (PIO):** You can team up with a local or regional media in order to give the ham radio the necessary publicity it deserves. The Public information officers are often the ones to foster communications between the community heads and some other organizations to bring about unity.

- **Official Observer (OO):** Official Observers assist other hams to avoid getting an FCC notice of rule violation due to operating or technical irregularities. They also ensure that there are no unlicensed

intruders or unauthorized transmission from unlicensed services.
- **Technical Specialist (TS):** If you possess useful skills in a particular area or if you are generally skilled in many parts of ham radio operations, you can opt to be a Technical Specialist. The Technical Specialist is like a consultant to other local and regional ham radio operators, as well as to the American Radio Relay League (ARRL). There are many other volunteering appointments you can know about from the ARRL's website at www.arrl.org/FandES/field/org.

ARES and RACES

The Amateur Radio Emergency Service (ARES) is controlled and managed by the American Radio Relay League Field Organization and works mainly with local public safety and NGO agencies, like the Red Cross. Local Amateur Radio Emergency Service (ARES) heads determine how best to task the volunteers and communicate with the agencies they are working with. You can register as a volunteer for ARES by filling the ARRL form obtainable as PDF on www.arrl.org/FandES/field/forms/fsd98.pdf. You will then mail the completed form to the ARRL. However, you will still have to join a local ARES team in order to be able to participate in training and exercises. The most convenient way to know more about the ARES organization within your locality is to communicate your ARRL Division's Section Manager (SM) by finding him or her on www.arrl.org/FandES/field/org/smlist.html. You can as well explore the American Radio Relay League's website on www.arrl.org/FandES/field/nets/client/netsearch.html. You can then find ARES networks in your locality.

The Radio Amateur Civil Emergency Service (RACES), initially established as a Civil Defense support club, is funded and controlled by the Federal Emergency Management Agency (FEMA) and is operated under the special FCC rules. RACES is organized and managed by a state or local civil defense agency charged with the responsibility for monitoring various disasters and provides ways of mitigating them.

This body is usually fully operational during disasters or emergencies when the states or federal officials heading it will usually call on the members for quick interventions. You can find more on the Radio Amateur Civil Emergency Service (RACES) on www.races.net.

If you are willing to join RACES, you can get started by searching for Auxiliary Communications Services (ACS) on your state government's website. The local ACS head is the one in charge of RACE membership in your area.

Preparing for an Emergency

Joining any of the emergency organizations mentioned above is actually a good way to get started. But this is not enough, you still need to acquaint yourself with the necessary emergency information to help you when any emergency surfaces. To get prepared for any emergency, you need to understand the following tips;
- Build relationships with other hams including the coordinators of the emergency group you have joined. Be familiar with the call signs of some of the local

clubs working from the government emergency operation centers (EOC). One way to get acquainted with the call sign is becoming a regular participant in many of the club's activities.
- You need to know where an emergency occurs. This is possible when you have the list detailing emergency net frequency together with the details of the leaders in each area.
- You need to know what an emergency situation is all about. You don't want to be caught between haste. Get your emergency kit ready to avoid getting yourself under pressure when an emergency situation eventually arises. This is necessary so you don't forget any important emergency kit at home during the emergency period. You can have a Go Kit which is just like your regular first-aid box for this purpose.
- You need to familiarize yourself with how to use any basic gadgets during an emergency situation. It is not enough to have a gadget, but knowing how to use such gadgets is the main skill. Spend more time getting trained in the procedures and

methods needed to get you ahead during an emergency. Be teachable and adaptable and watch your productivity soars.

Tips to get ahead during emergency

1. Your first responsibility is to your family. Endeavor to check first, during an emergency, that you and the rest of your families are safe and secured. After your safety, you can then respond as an emcomm volunteer.
2. Have control of your main emergency frequencies.
3. Adhere to the instructions you get from the network control or any other emergency official on your frequency. Endeavor to know if and when check-ins are requested.
4. Seek your local emergency communications head or designee for any other instructions.

Reporting an accident: Tips

1. Turn up your ham radio's power very high and start saying, "Break" or
"Break Emergency" at the very first opportunity you see.

This is necessary provided that one station is weak, a stronger signal will get you the listening ears of other available stations. Don't worry about interrupting an ongoing chat, of course you are in an emergency situation.
2. Once you have a hold on the frequency, intimate the station that you are in an emergency situation.
3. Tell them very clearly that you are rendering an emergency autopatch and then turn on the autopatch system. If you are having issues turning on your repeater's autopatch, try and contact another repeater to help you with it. Alternatively, you can put on your VHF or HF and ask anyone to intimate 911 of pending emergency. Tell them all the important materials and don't leave your frequency until you get a response that your call is complete and your message has been sent.
4. Call 911 and when the operator picks, tell your name and intimate that you want to report an emergency through the amateur radio.

5. Follow the guides of the operator from that point. If the operator tells you to stay on the line, kindly oblige and tell the other user to also stay put.
6. As soon as the operator finishes, release your autopatch and tell them that you have released the autopatch.

If you want to report any automobile accident, for instance, you should know:
_ The street number of the highway number.
_ The address of the street or highway.
_ The direction or route where the accident took place.
_ If the accident is actually blocking vehicular traffic
_ If there are any injuries or deaths.
_ If the vehicles erupt fire, smoke, or have spilled fuel on the road.

CHAPTER FIVE

MAINTAINING YOUR STATION

Tools for maintaining your equipment

- **Wire cutters:** For cutting either small or big wires.
- **Soldering iron and gun:** You can get a model that you will be able to adjust its temperature. Most connectors and circuitboards require a low temperature soldering iron with a sharp tip.
- **Terminal crimpers:** this works in place of pliers. Use it on crimp terminals to navigate any pull out connections or loose connections.
- **Head-mounted magnifier:** Enable you to see even the smallest of electronic components that you might be working with.
- **Volt-ohmmeter (VOM):** If you have the money, endeavor to buy one that has diode and transistor checking, continuity tester, and probably a capacitor and inductance checker.

Together with adapters and your spare parts, you should have ready some consumable parts;
- **Fuses:** Have some spare fuses ready.
- **Electrical tape:**
- **Fasteners:** For fastening purposes.
- **Interference suppressors:** Having some filters and ferrite cores around enables you to solve interference issues quickly.

Cleaning equipment is a very essential part of maintenance tools and you may probably need the following items:
- **Soft bristle brushes:** Small Old paint brushes and toothbrushes are particularly excellent cleaning tools. You can get inside gadgets holes with a small brush.
- **Metal bristle brushes:** Light steel and brass brushes help clean up oxide and corrosion. Brass brushes do not destroy metal connectors, but can damage plastic knobs or plastic displays. Don't forget to clean the brush you used for corrosion or grease cleaning after use.
- **Solvents and sprays:** You can get isopropyl alcohol, compressed gases or contact cleaner. These help clean cabinets and panels.

Repairing and building tools

Consider the following handy specialty tools and gadgets for simple adjustments and measurements;
- **Wattmeter or SWR meter:** This is a power measurement device for troubleshooting. The Bird Model Wattmeter is advisable as it is proven and tested. You can also get some inexpensive models depending on your budget. Avoid buying wattmeters not properly calibrated.
- **RF and audio generators and oscilloscope:** Radios have a lot to do with alternating current (AC) signals and you require a device to generate and access these signals. Although, a DC test and a voltmeter can also help you with some tasks involving Direct current aspect, but radio signals are of AC. You can navigate the American Radio Relay League (ARRL) website on www.arrl.org/tis/info/HTML/Hands-On-Radio/.to get and choose your electronics.
- **Nibbling tool and chassis punch:** Starting from a round hole, the nibbler is esse-

ntially a hand-held punch that chops out a little rectangle of sheet metal or any plastic. You can use a nibbler to cut out a large rectangular or any irregular opening and then make the hole to shape. The chassis punch can make a sizable hole in up to one-eight inch aluminum or some 20-gauge steels.
- **T-handled reamer and countersink:** The reamer gives you a chance to enlarge a small hole to give you the size you want.

Routine maintenance for stations

- Ensure that all connectors are strong. This is because heat generated within might have worked them loose. Check the feedlines for any damages.
- **Test transmitters and amplifiers for full power output on all bands:** Endeavor to check the antenna and the RF cables.
- **Check received noise level on all bands:** The noise level will mostly tell you whether your feedlines are in good working condition.

- **Check SWR on all antennas:** A small change in the frequency of the SWR could mean a connection problem.
- **Check all antennas and outside the feedlines:** You can deploy a pair of binoculars or climb up and inspect. Look for loose connections and fix them.
- **Check masts, antenna stands and towers:** This is necessary to check if weather has damaged some part so that you can easily fix them.
- **Clean your operating table frequently:** You don't want to have a messy desk as this reduces concentration.

ABOUT AUTHOR

Colin Schmitt is a renowned radio personality who has dedicated over 15 years of his life doing ham radio as a hobby. Robert enjoys teaching people and is always happy giving people all that he knows about amateur radio. Robert has participated in many emergency operations as hams where he got to display his exceptional radio skills.

Robert lives in Minnesota, United States. He is happily married with two beautiful daughters.

www.ingramcontent.com/pod-product-compliance
Lightning Source LLC
Chambersburg PA
CBHW050011230526
45465CB00003BB/1373